A Journey Through
Your Amazing
Heart

WRITTEN & ILLUSTRATED BY

RAYAAN SIDDIQUI

Printed in the United States of America

First Printing, 2022

ISBN-13: 978-0-578-98847-4

About the Author

Rayaan Siddiqui produced his first book, *A Journey Through Your Amazing Brain and Senses*, after he was inspired by what he learned at Brown University's Pre-College neurology program. He has always been fond of art and enjoys oil painting as a hobby.

Recognizing that children were interested in learning more about the human body, he turned his attention to researching another key organ – the heart. He has coupled his artistic abilities and his inquisitive nature to create yet another book.

With this easy-to-understand book and simple illustrations, he hopes that he can inspire inquisitive young minds to learn about the complexities of the heart and its amazing power within the human body.

Currently, Rayaan is a junior in high school and resides in Doha, Qatar with his family and aims to expand his knowledge about medicine in the future.

Acknowledgments

They say that success breeds success. In my case, it has led to increased curiosity. After getting engrossed in the inner workings of the brain and its relationship with the senses, I got hooked onto the heart. After all, one had to consider about which has greater power – the heart or the mind.

Writing this second book has been equally as challenging and rewarding as the first one. And on this journey, I would like to thank my dad, Kamran, for helping to finance this project. Of course, this second book wouldn't have been possible without my mom, Nadia, who continued to devote her time as I ran my research and illustration ideas by her. I would also like to thank my younger sister, Zoha, who was a great sounding board throughout this project.

Finally, I am truly grateful to Linda Hargrove and her expertise in stitching my artwork and content together in this book, as well as providing editorial insights.

The Amazing Heart

The heart is amazing. A human's heart is located inside the ribcage, in the middle, tilted slightly to the left. The primary role of the heart is to pump blood around the body.

The blood that is pumped by the heart contains nutrients, oxygen, and hormones **to be transported to every part of the body**.

Take your right hand and put it on the left side of your chest. Can you feel your heartbeat? That's amazing!

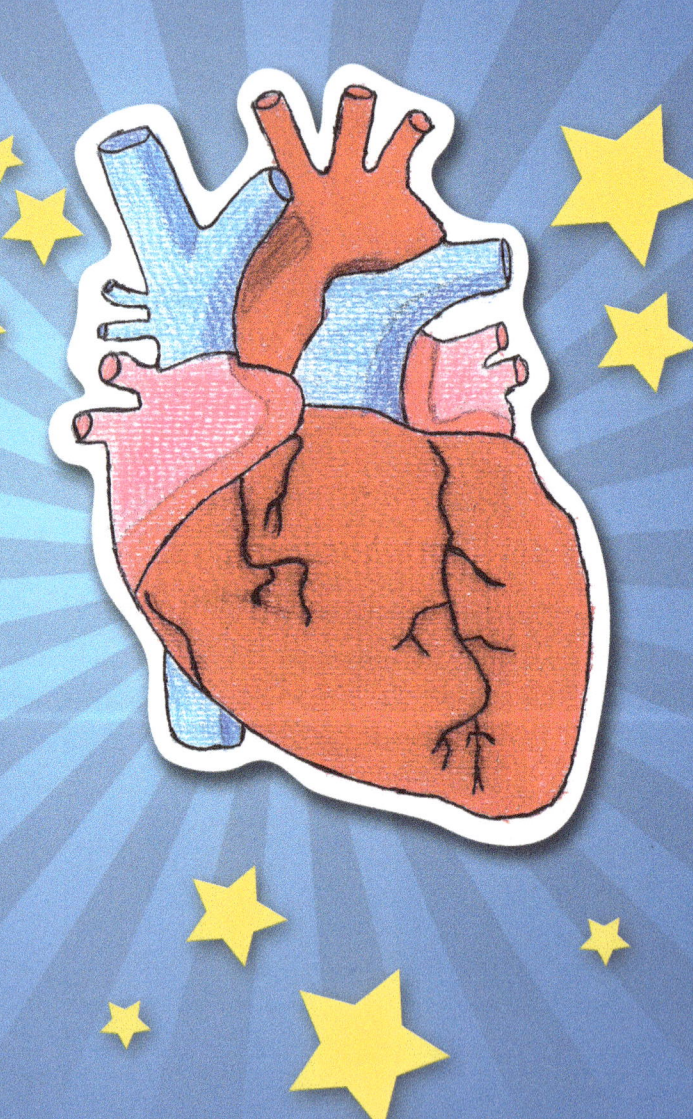

Inside the Heart

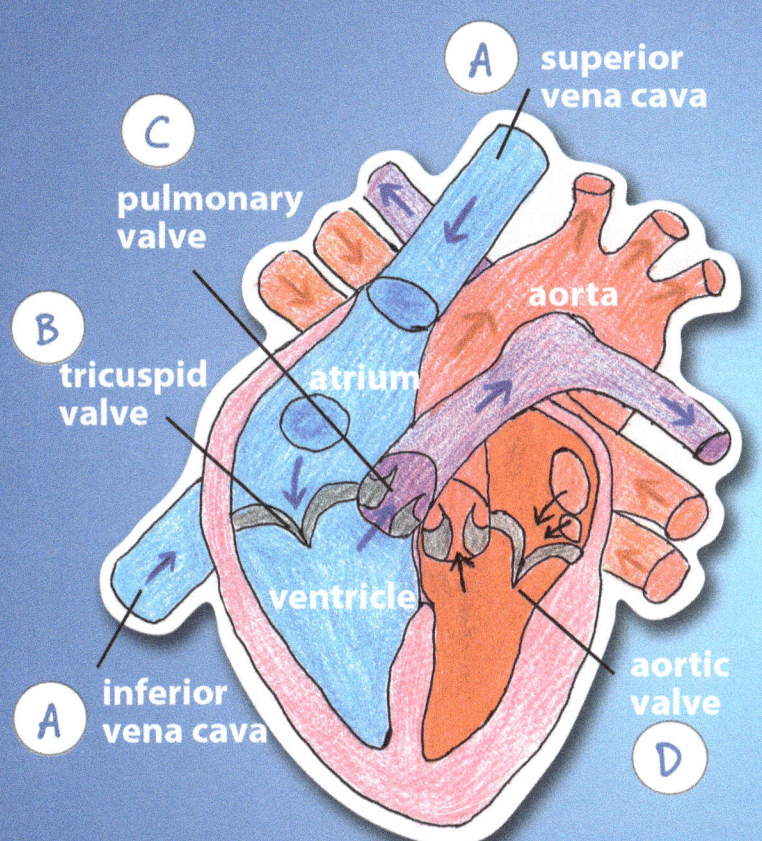

A superior vena cava

C pulmonary valve

B tricuspid valve

atrium

aorta

ventricle

A inferior vena cava

aortic valve

D

Blood enters the heart (A) and travels through (B). Then it goes to (C) and heads toward the lungs to get oxygenated.

The lungs help to remove carbon dioxide (CO_2) and provide the blood with oxygen (O_2). The oxygenated blood from the lungs goes back into the heart, into the left atrium, then to the left ventricle, passing through (D).

Finally, the blood travels to the aorta where it is sent throughout the body.

Valves are like doors with tight seals. They make sure that the blood does not go backwards.

Did you know that the heart is roughly the size of your clenched fist?

Blood Circulation

The left side of your heart sends *oxygen-rich* blood throughout your body to be used. While the oxygen is being used, carbon dioxide (CO_2) gets carried away. Then, this *oxygen-poor* blood comes back to the right side of the heart.

The **right ventricle** pumps the blood to the lungs to pick up oxygen (O_2). In the lungs, CO_2 is removed from the blood and sent out of the body when we exhale. After that, we inhale and the process starts all over again.

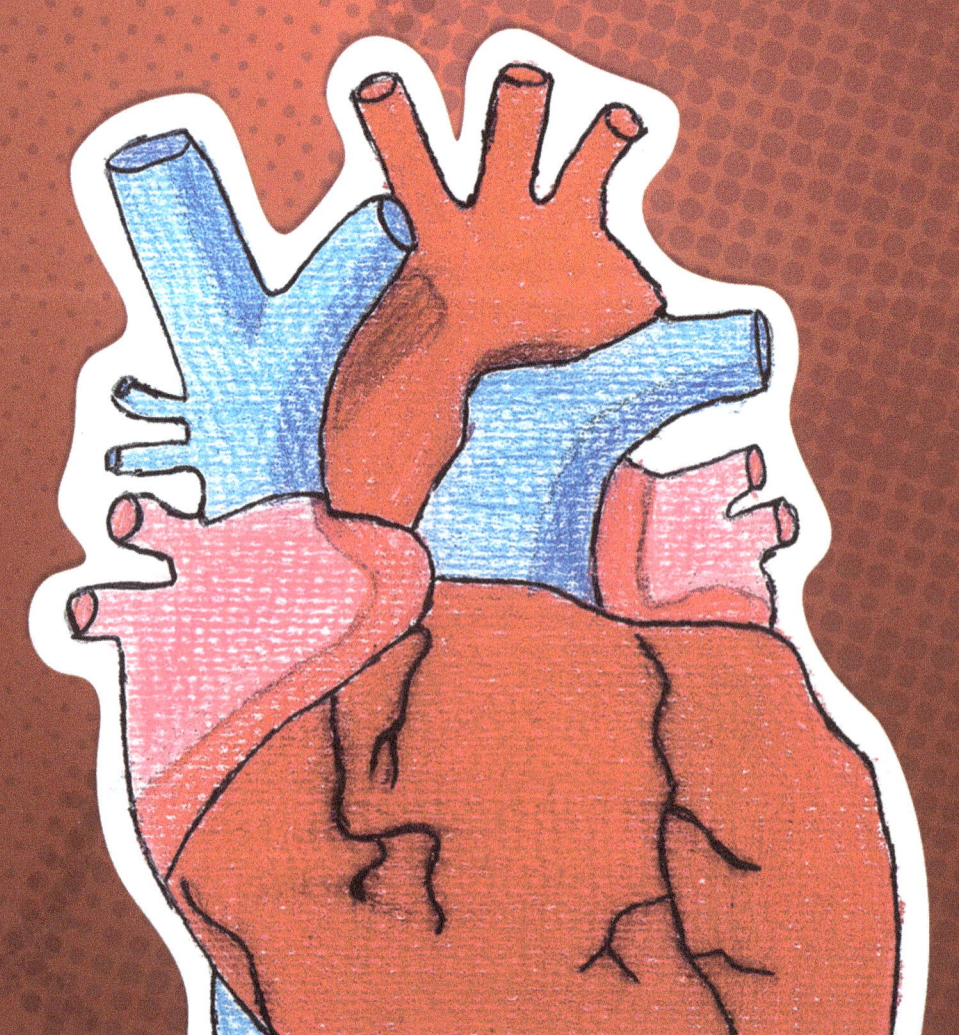

3

Blood Circulation

(Continued)

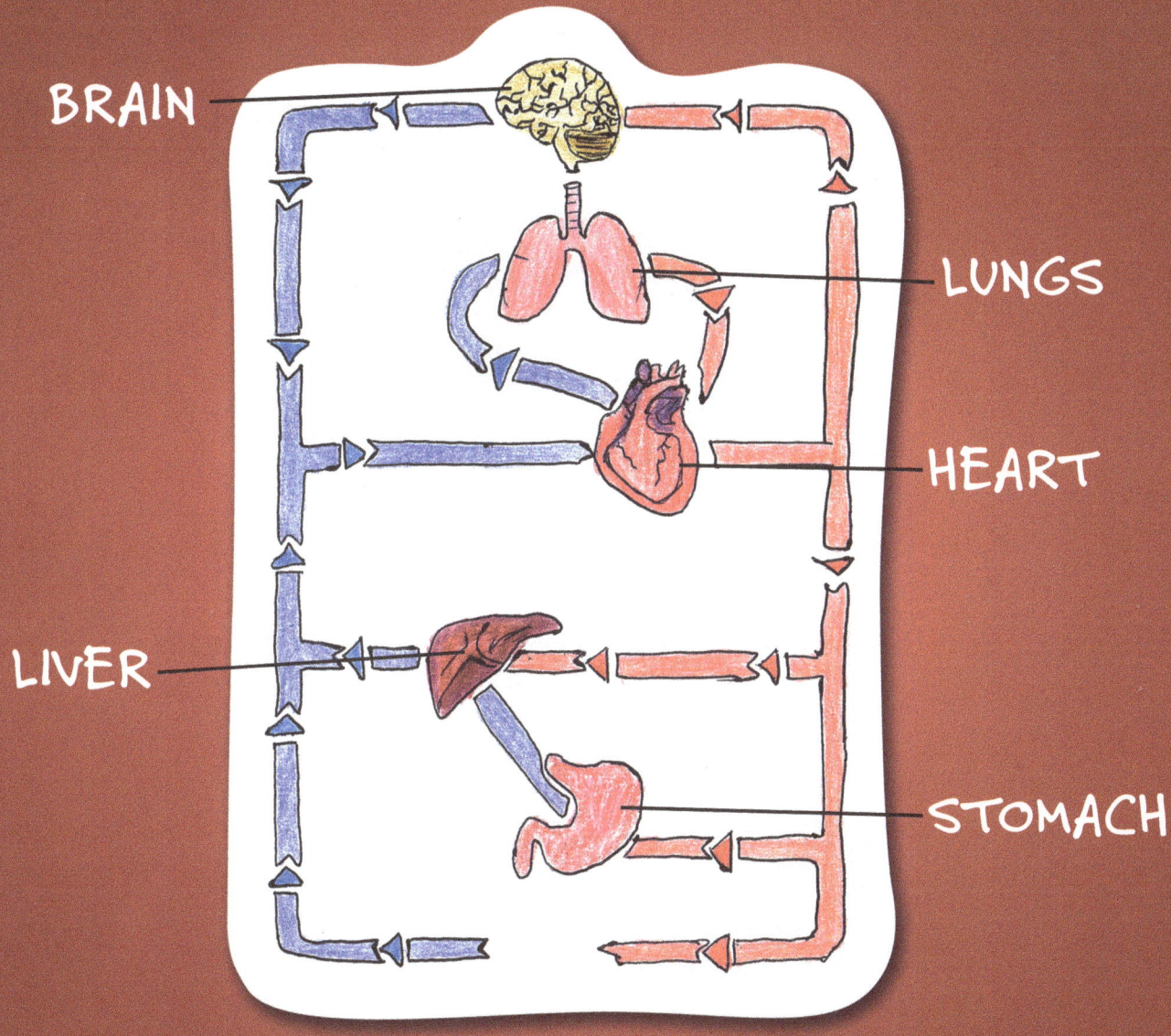

BRAIN

LUNGS

HEART

LIVER

STOMACH

After the blood passes through the heart, it needs to get to your body. Instead of just sloshing around inside you, it's transported through vessels called arteries and veins. Blood vessels that carry blood to the heart are called veins and vessels that carry blood away from the heart are called arteries. This constant process is called the **circulatory system** and can be completed in less than 60 seconds.

Strengthening the Miracle Muscle

Your heart is a muscle. Like any other muscle in your body, you have to exercise and eat healthy to keep your heart strong.

As you exercise more frequently, your heart becomes more efficient at pumping blood throughout your body. This means that your heart can pump more blood in just a single beat. This is known as the stroke volume. This allows the heart to beat slower, therefore keeping your blood pressure under control.

The heart pumps between 1,500 and 2,000 gallons of blood every day. That's equivalent to 24,000 and 32,000 glasses of orange juice every day!

Feeding the Miracle Muscle

Eating healthy foods is just as important as exercising. Fast food has been associated with poor diet. These foods have higher saturated fats and more sugar which can cause heart disease. Excess salt in foods like burgers and fries are known to cause high blood pressure which can lead to heart disease and stroke. A low saturated fat, high-fiber diet that is high in plant-based foods can reduce the chances of developing heart disease.

HEALTHY

UNHEALTHY

Heart Disease
& Other Problems

There are some common heart diseases that can be caused by the environmental influences like cigarette smoke. Smoking increases your heart rate and tightens your arteries, forcing your heart to work harder than it really should. Smoking also causes irregular heartbeats and raises blood pressure, which is one of the main causes of stroke. It can be fatal.

Did you know that a heart attack can also be caused by high levels of stress? In a heart attack, there is a loss of blood supply, mainly when there is a clot in one of the arteries.

Cardiac arrest is slightly different from a heart attack. In cardiac arrest, there is an electrical malfunction in the heart that causes an irregular heartbeat. This can also be fatal.

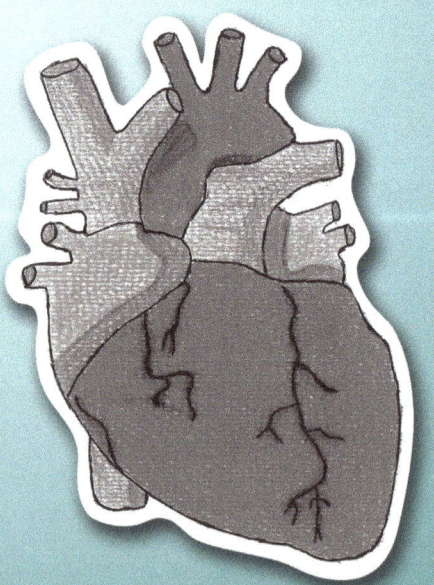

Heart Disease
& Other Problems

(Continued)

As you grow older, your heart has to work harder and sometimes it can't keep up. This is why it's important to stay active especially as you age. This will prevent age-related heart disease.

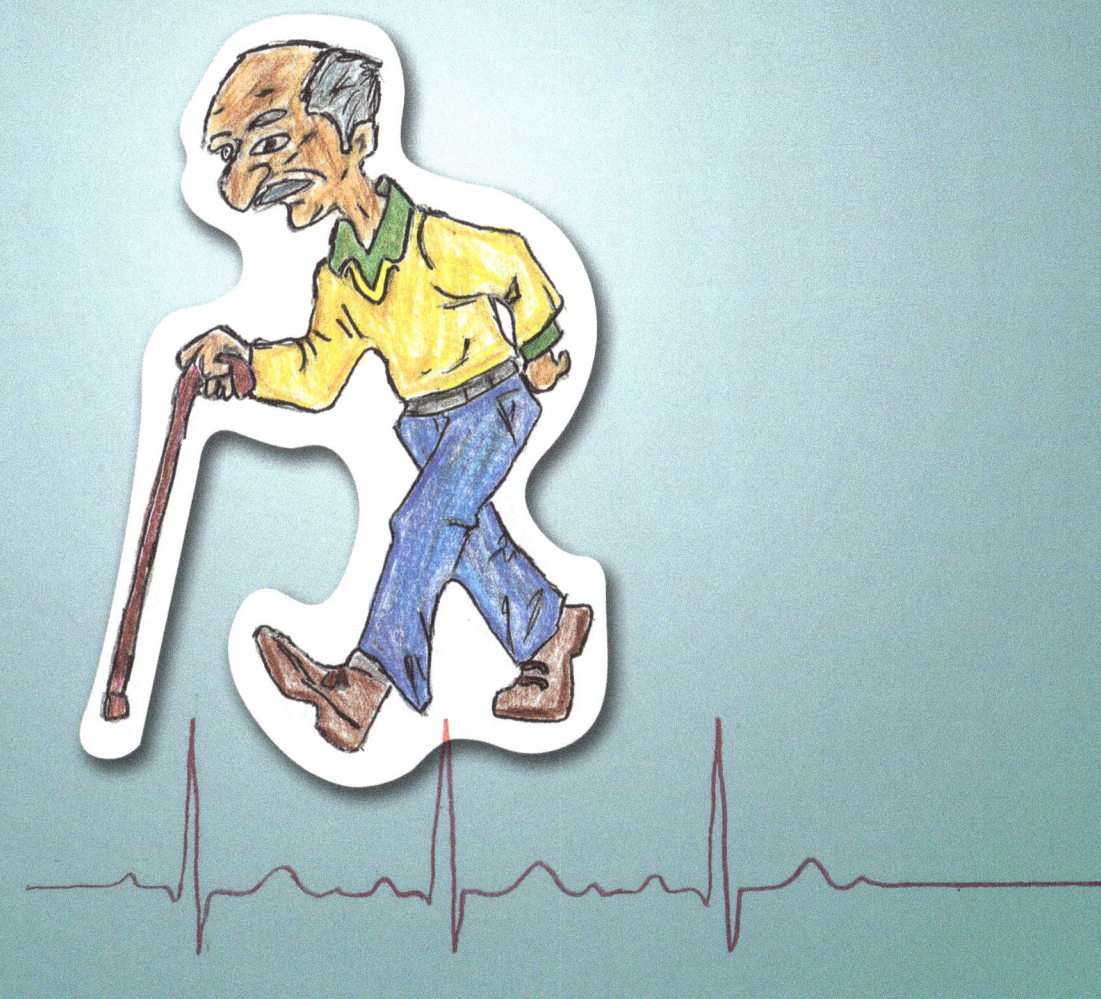

Arteries, Veins, Capillaries

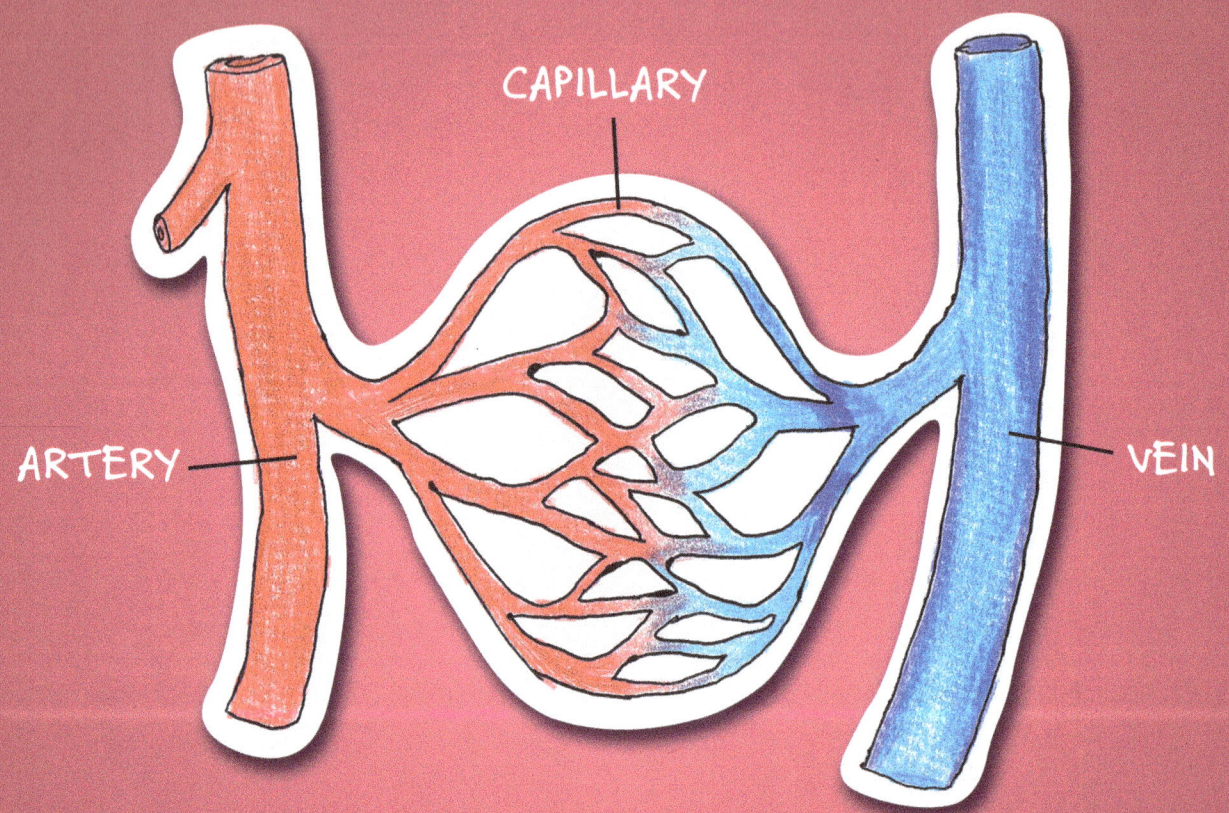

Arteries take oxygenated blood away from the heart.

Veins take oxygen-poor blood toward the heart to be oxygenated.

Capillaries are the smallest blood vessels and distribute oxygenated blood from arteries to the tissues of the body and then take deoxygenated blood from the tissues back in the veins.

Keep It Pumpin'

Contrary to popular belief, the heart is not actually on the left side of your chest.

Your heart is situated exactly in the middle but the way that it is positioned makes it pump toward the left.

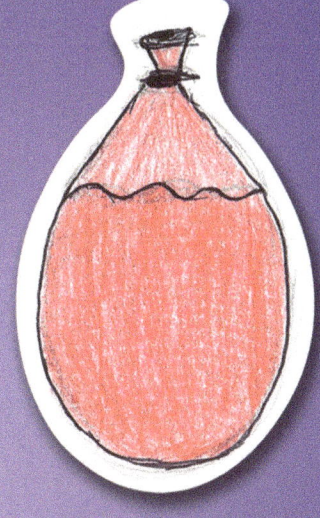

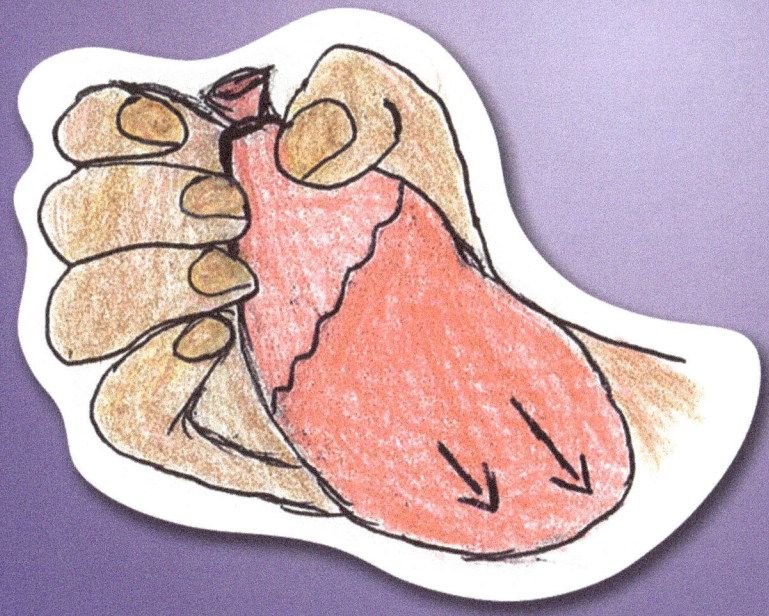

Imagine holding a water balloon (in this case, it's your heart). When the balloon is squeezed, the heart contracts and pumps outward, on the left side and that is how the heart would look, when pumping blood.

It's Electrifying!

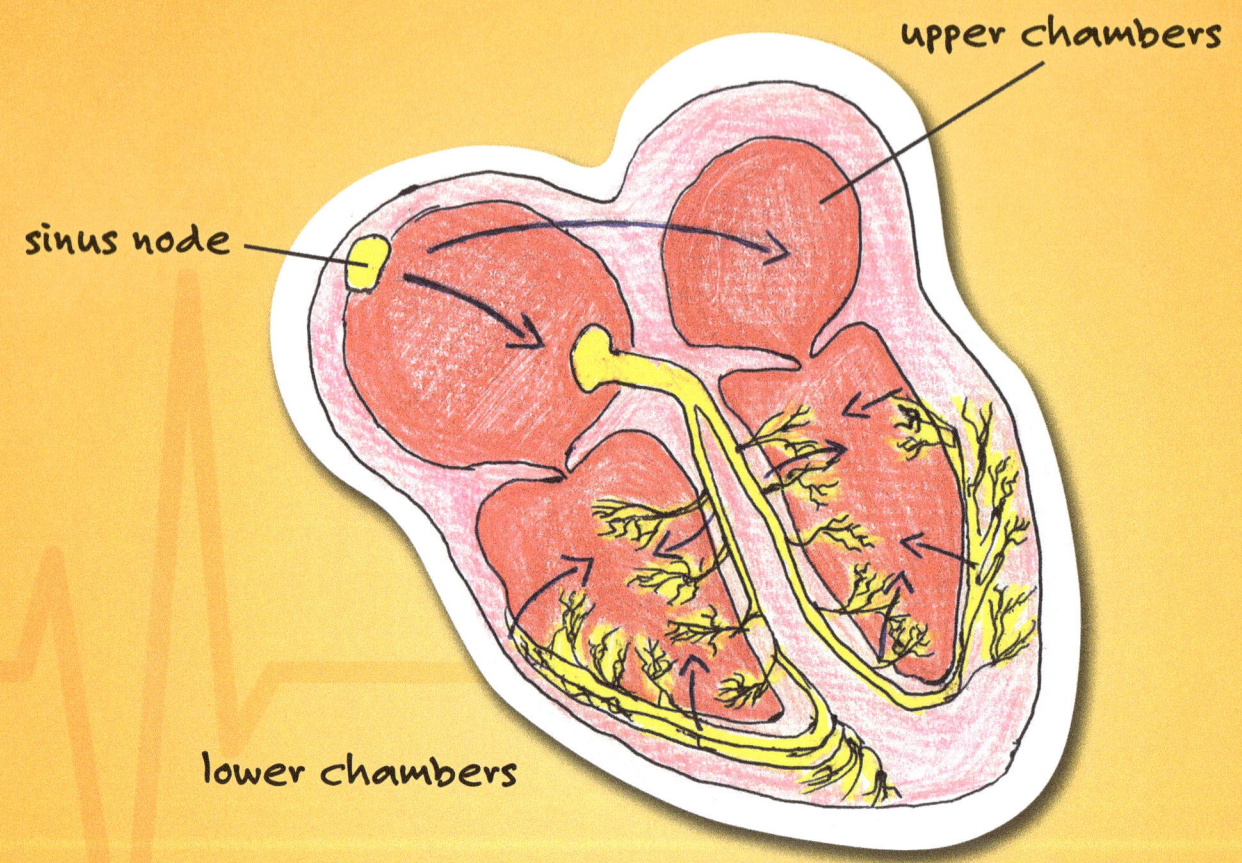

upper chambers

sinus node

lower chambers

In order for the heart to squeeze and pump blood, it needs an electrical impulse to start a heartbeat. The electrical impulse starts on the right side of the upper chamber in an area called the sinus node. The sinus node is the heart's natural pacemaker.

The impulse leaves the sinus node and travels down a path through the upper chambers (the atria), causing them to contract and squeeze blood into the lower chambers. This creates a heartbeat.

That one signal can happen over and over **115,200** times a day!

Checking ...
Checking...

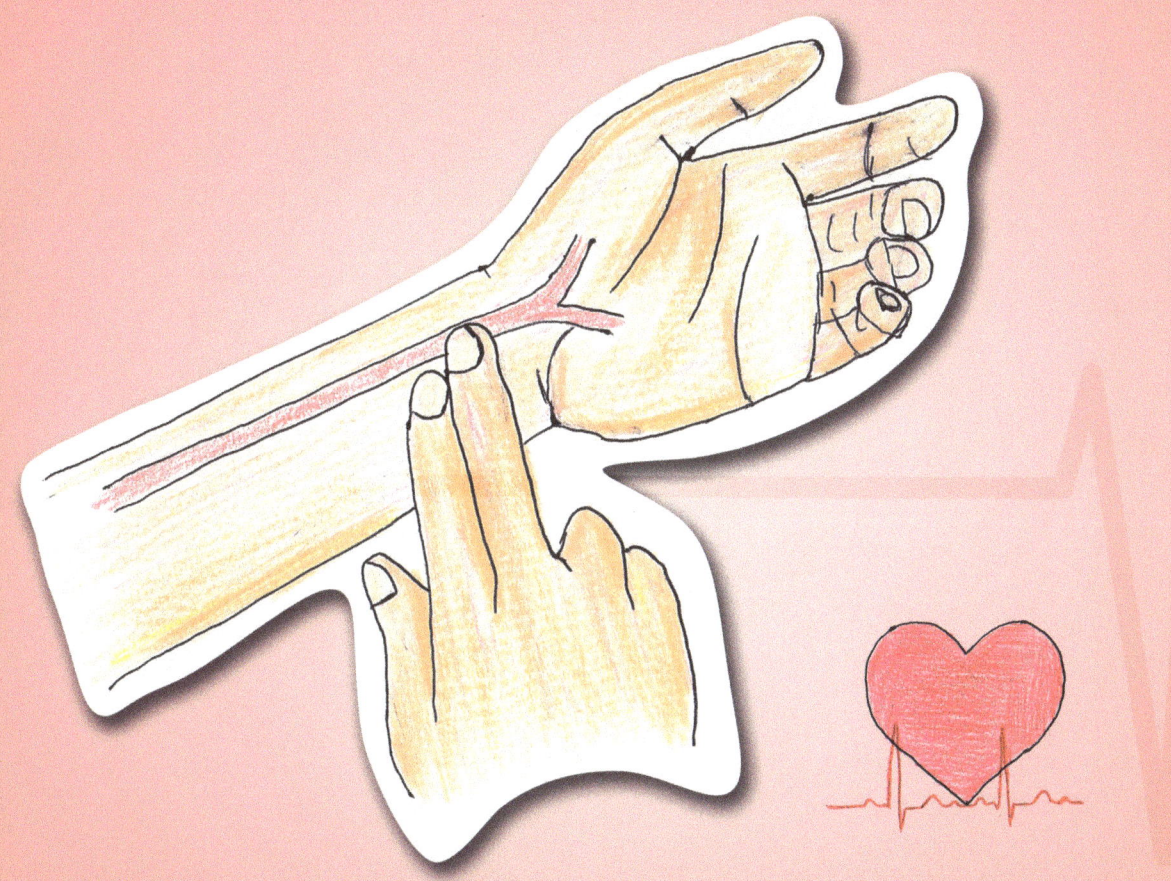

Even though your heart is inside you, there is still a way to know it's working from the outside. It's your pulse. You can find your pulse by lightly pressing on the skin where there's a large artery. Two good places to find it are on the side of your neck and on the underside of your wrist.

You'll know that you've found your pulse when you can feel a small beat under your skin. Each beat is caused by the contraction (squeezing) of your heart. If you want to find out what your heart rate is, go to the experiments page at the end of this book. When you are resting, you will probably feel between 70 and 100 beats per minute.

Animals with Hearts

Some animals have more than one heart as a way to get blood pumping around the body faster because of their active lifestyles or to get blood around different areas of the body. Sometimes, one heart is too weak to pump blood on its own and needs more than one heart.

Octopus

An octopus has a very active lifestyle and cannot just rely on one heart. Octopuses have three hearts. One to pump blood around the body and two other hearts to pump blood to the gills. Octopus also have a slightly different, less efficient blood than humans so the three hearts compensate for that.

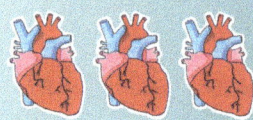

Hagfish

Hagfish, a type of marine fish, have four hearts (which are quite a lot for its size - reaching a maximum length of just 20 inches). One heart serves as a primary pump while the other three are accessory pumps.

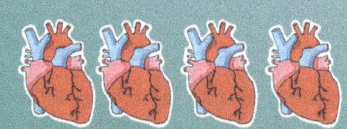

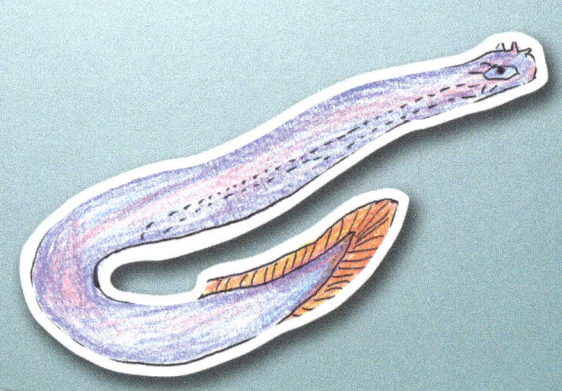

 Blue whales have the largest hearts of any mammal!

Animals With Hearts

(Continued)

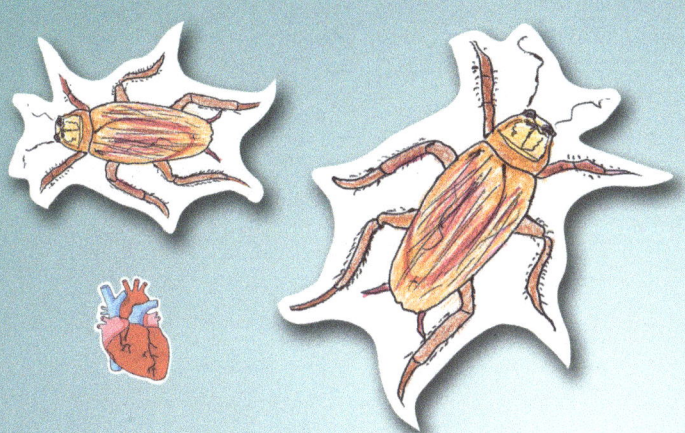

Cockroach
Cockroaches just have one heart but have 13 chambers. Humans, which are much larger, have one heart with four chambers. In case of a failure of one of the chambers in the cockroach's heart, it does not become a life-threatening issue and can remain alive.

Snake
Snakes also just have one heart but only have three chambers. Since snakes are reptiles, they are cold-blooded and the three chambers allow for better separation of oxygenated and deoxygenated blood.

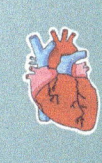

Squid
Like octopuses, squids also have three hearts. Again, they have two hearts that pump blood to the gills called branchial hearts and one main heart that pumps blood to the rest of the body, called the systematic heart. Similar to an octopus, three hearts are needed to support the active lifestyle.

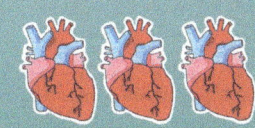

Lub Dub "Experiment"

When you go to the doctor's office, your pulse will be checked using an instrument called a **stethoscope**. Even if you don't have one of these at home, you can use your fingers to find your pulse. The aim of this experiment is to see how much higher your heart rate rises when doing exercise.

1. Take your pointer finger and middle finger and place them on the side of your neck. Try and locate a beat.
2. If you cannot find a beat on the side of your neck, try placing your fingers on the inside of your wrist.
3. Once you've found the beating sound, start a timer for 60 seconds and see how many beats you count, when resting, it should be around 60 - 100bpm.
4. Now, stand up and do an exercise such as high knees or heel flicks for 1 minute.
5. Quickly, before your heart rate drops, find your heart beat and repeat step 3. However, this time, your heart rate should be higher than what your **resting heart rate was.**

The heart rate is higher because more oxygen needs to be delivered at a faster rate.

Lightning Source UK Ltd.
Milton Keynes UK
UKHW050609020323
417740UK00016B/170